Extreme Weather

HURRICANES

Martha London

DiscoverRoo
An Imprint of Pop!
popbooksonline.com

abdobooks.com

Published by Pop!, a division of ABDO, PO Box 398166, Minneapolis, Minnesota 55439.

Printed in the United States of America, North Mankato, Minnesota.

052019
092019

THIS BOOK CONTAINS RECYCLED MATERIALS

Cover Photo: Shutterstock Images
Interior Photos: Shutterstock Images, 1, 17, 22–23; iStockphoto, 5, 11, 16, 18, 20–21, 25, 26, 28, 29 (hurricane), 29 (checkmarks), 29 (car), 29 (food), 29 (window), 29 (television), 29 (bottom), 31; NOAA, 6, 7, 8, 9, 14–15, 30; Nicolas Marino/ Novarc Images/Alamy, 13; David Wall/Alamy, 19; Bettmann/ Getty Images, 27

Editor: Connor Stratton
Series Designer: Sophie Geister-Jones

Library of Congress Control Number: 2018965058

Publisher's Cataloging-in-Publication Data

Names: London, Martha, author.

Title: Hurricanes / by Martha London.

Description: Minneapolis, Minnesota : Pop!, 2020 | Series: Extreme weather | Includes online resources and index.

Identifiers: ISBN 9781532163944 (lib. bdg.) | ISBN 9781532165382 (ebook)

Subjects: LCSH: Hurricane waves--Juvenile literature. | Storm surges--Juvenile literature. | Severe storms--Juvenile literature. | Natural disasters--Juvenile literature.

Classification: DDC 551.552--dc23

WELCOME TO DiscoverRoo!

Pop open this book and you'll find QR codes loaded with information, so you can learn even more!

Scan this code* and others like it while you read, or visit the website below to make this book pop!

popbooksonline.com/hurricanes

*Scanning QR codes requires a web-enabled smart device with a QR code reader app and a camera.

TABLE OF CONTENTS

CHAPTER 1

HOW HURRICANES FORM

A hurricane is a huge storm with strong winds. The winds blow at least 74 miles per hour. The storm swirls around like a spinning top. But the center is calm. This part is called the eye.

WATCH A VIDEO HERE!

Most hurricane eyes are between 20 and 40 miles wide.

On August 30, 2018, a tropical disturbance formed over the Atlantic Ocean.

Hurricanes begin over the ocean. Warm air causes many storm clouds to form over a **tropical** area. Together, these clouds are called

By August 31, 2018, the disturbance had become a tropical depression.

a tropical **disturbance**. As the disturbance grows, the clouds begin to spin. When that happens, the storm is called a tropical depression.

The storm gains strength as it moves across the water. Its winds pick up. If winds reach 39 miles per hour, the depression turns into a tropical storm.

By September 1, 2018, the depression had grown to a tropical storm.

By September 4, 2018, the tropical storm had grown to a hurricane.

A tropical storm becomes a hurricane when its winds reach at least 74 miles per hour.

DID YOU KNOW?

Scientists give each tropical storm a name.

CHAPTER 2

WHERE HURRICANES FORM

Hurricanes happen all over the world. Most of them happen around the **equator**. Water is warmer near the equator. The heat rising from the water builds bigger storms.

LEARN MORE HERE!

Earth's equator cuts across the middle of Africa.

DID YOU KNOW?

More than half of all hurricanes happen in August and September.

Many storms begin off the coast of Africa. Warm air over the Sahara Desert mixes with cooler, more humid air from the coast of central Africa. The warm air rises and starts to spin. The humid air **condenses** into water and forms clouds. These clouds also spin. The spinning clouds can become hurricanes.

ONE STORM, MANY NAMES

Hurricanes are a type of tropical **cyclone**. Around the world, people call these storms different names. Tropical cyclones that hit North America and the Caribbean are called hurricanes. In Asia, they are called typhoons. South of the equator, these storms are called cyclones.

The Sahara Desert stretches approximately 3,000 miles from Egypt to the Atlantic Ocean.

The United States is not close to the equator. But hurricanes still hit the country. A warm **current** of water

A hurricane travels across the Gulf of Mexico.

flows into the Gulf of Mexico from the south. This current is perfect for making big storms.

CHAPTER 3

DAMAGE

Hurricane Harvey hit the US coast in August 2017. Houston, Texas, received more than 4 feet of rain. Within a week, one-third of the city had flooded. Approximately 39,000 people had to **evacuate**.

Many parts of Houston, Texas, remained flooded long after Hurricane Harvey ended.

LEARN MORE HERE!

A single hurricane can dump enough rain to cover all of Texas with 4 inches of water. When a hurricane hits land, there is also a **storm surge**. This surge causes flooding along the coast. The ocean can rise up to 20 feet.

A storm surge hits a town in New Zealand.

DID YOU KNOW?

Nine out of ten deaths caused by hurricanes are the result of storm surges.

The most powerful hurricanes

have winds faster than a roller coaster.

The wind can pull roofs off houses.

A hurricane can cause a house to fall apart.

It can uproot trees and fences. These trees and fences can hit other houses and buildings, causing more damage.

THE FIVE CATEGORIES OF HURRICANES

Scientists rank every hurricane into a category. These categories tell people how strong each hurricane is. The categories are based on the power of a hurricane's winds.

1 Minor Damage
winds of 74 to 95 miles per hour

2 Medium Damage
winds of 96 to 110 miles per hour

3
Lots of Damage
winds of 111 to 129 miles per hour

4
Extreme Damage
winds of 130 to 156 miles per hour

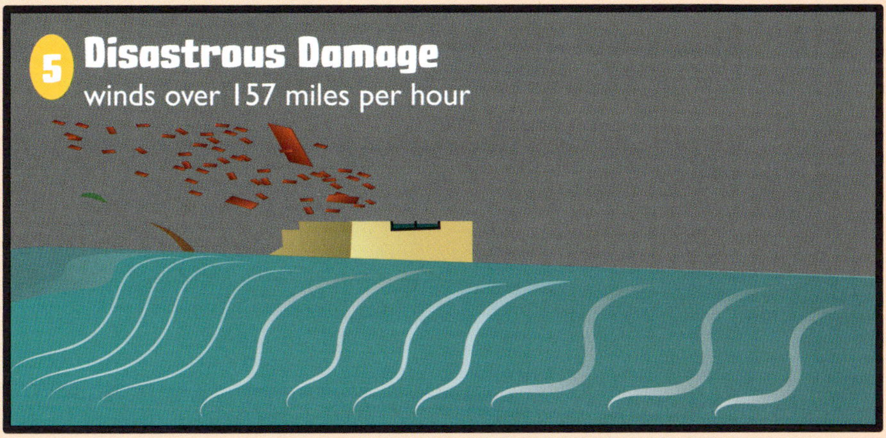
5
Disastrous Damage
winds over 157 miles per hour

CHAPTER 4

STAYING SAFE

Scientists can track where a hurricane will go a few days ahead of time. When a hurricane is very strong, people may be told to **evacuate**. They will need to leave the area.

COMPLETE AN ACTIVITY HERE!

People evacuate in their cars before a storm comes.

Before leaving, people protect their homes and businesses in many ways. They use wood panels to board their windows. These panels protect the glass from flying objects during the storm.

DID YOU KNOW?

The eye of a hurricane is calm. But the storm might not be over. People should wait until officials announce the danger has passed.

People board up a store to protect its windows before a hurricane.

In 2012, New York City used sandbags to protect its subway stations from Hurricane Sandy.

Some people pile sandbags for protection from floods. They also place tarps. Sandbags and tarps help keep water out. If people prepare, they can reduce a hurricane's damage. People can also keep themselves safe.

Hurricane Safety Checklist

- ✓ Evacuate if necessary.
- ✓ Keep enough fresh water and food for three days.
- ✓ Stay away from windows during the storm.
- ✓ Follow the news for weather updates.

MAKING CONNECTIONS

TEXT-TO-SELF

Are hurricanes common where you live? Why or why not?

TEXT-TO-TEXT

Have you read books about other kinds of storms? What do they have in common with hurricanes? How are they different?

TEXT-TO-WORLD

Hurricanes bring strong wind and rain. What might people have to fix after a hurricane hits?

GLOSSARY

cyclone – a storm that rotates and often brings rain.

disturbance – a change in normal weather conditions.

equator – an imaginary line that runs around the middle of Earth.

evacuate – to leave a place in order to stay safe.

storm surge – a rise in ocean levels that causes flooding.

tropical – having to do with a place where the weather is usually warm and wet.

INDEX

ONLINE RESOURCES

popbooksonline.com

Scan this code* and others like it while you read, or visit the website below to make this book pop!

popbooksonline.com/hurricanes

*Scanning QR codes requires a web-enabled smart device with a QR code reader app and a camera.